EMBRYOLOGY PROCEDURE LOG BOOK

PROCEDURE	NOTES
DAILY QC	
MEDIA/DISH PREP	
OOCYTE RETRIEVAL	
SPERM PROCESSING	
SPERM CRYOPRESERVATION	
SPERM THAW	
OOCYTE INSEMINATION	
INTRACYTOPLASMIC SPERM INJECTION	
FERTILIZATION ASSESSMENT	
DAY THREE EMBRYO GRADING/ AH	
BLASTOCYST GRADING	
EMBRYO TRANSFER	
EMBRYO CRYOPRESERVATION	
EMBRYO THAW	

Quality Control Procedure Notes

Goals and Expectations

Media and Dish Prep Procedure Notes

Goals and Expectations

Oocytes Retrieval Procedure Notes

Goals and Expectations

Oocytes Retrieval Cases

Date	Patient Initials	Assisted or Solo	# of Oocytes Retrieved	# of Oocytes Missed	Stripping	# MII	# MI	# GV	# DEG	# EZ	Notes

Oocytes Retrieval Cases

Date	Patient Initals	Assisted or Solo	# of Oocytes Retrieved	# of Oocytes Missed	Stripping	# MII	#MI	# GV	# DEG	# EZ	Notes

Oocytes Retrieval Cases

Date	Patient Initials	Assisted or Solo	# of Oocytes Retrieved	# of Oocytes Missed	Shipping	# MII	# MI	# GV	# DEG	# EZ	Notes

Oocytes Retrieval Cases

Date	Patient Initials	Assisted or Sole	# of Oocytes Retrieved	# of Oocytes Missed	Stripping	# MII	# MI	# GV	# DEG	# EZ	Notes

Oocytes Retrieval Cases

Date	Patient Initials	Assisted or Solo	# of Oocytes Retrieved	# of Oocytes Missed	Stripping	# MII	# MI	# GV	# DEG	# EZ	Notes

Oocytes Retrieval Cases

Date	Patient Initials	Assisted or Solo	# of Oocytes Retrieved	# of Oocytes Missed	Stripping	# MII	# MI	# GV	# DEG	# EZ	Notes

Oocytes Retrieval Cases

Date	Patient Initials	Assisted or Solo	# of Oocytes Retrieved	# of Oocytes Missed	Stripping	# MII	# MI	# GV	# DEG	# EZ	Notes

| Date | Patient Initials | Assisted or Solo | # of Oocytes Retrieved | # of Oocytes Missed | Stripping | # MII | # MI | # GV | # DEG | # EZ | Notes |

Oocytes Retrieval Cases

Date	Patient Initals	Assisted or Solo	# of Oocytes Retrieved	# of Oocytes Missed	Stripping	# MII	#MI	# GV	# DEG	# EZ	Notes
Date	Patient Initals	Assisted or Solo	# of Oocytes Retrieved	# of Oocytes Missed	Stripping	# MII	#MI	# GV	# DEG	# EZ	Notes

Sperm Processing Procedure Notes

Goals and Expectations

Sperm Processing Cases

Date	Patient Initials	Pre Process Volume	Pre Process Motility	Type of Process	Post Process Volume	Post Process Motility	Notes

| Date | Patient Initials | Pre Process Volume | Pre Process Motility | Type of Process | Post Process Volume | Post Process Motility | Notes |

Sperm Processing Cases

Date	Patient Initials	Pre Process Volume	Pre Process Motility	Type of Process	Post Process Volume	Post Process Motility	Notes

Sperm Processing Cases

Date	Patient Initials	Pre Process Volume	Pre Process Motility	Type of Process	Post Process Volume	Post Process Motility	Notes

Sperm Processing Cases

Date	Patient Initials	Pre Process Volume	Pre Process Motility	Type of Process	Post Process Volume	Post Process Motility	Notes
Date	Patient Initials	Pre Process Volume	Pre Process Motility	Type of Process	Post Process Volume	Post Process Motility	Notes

Sperm Processing Cases

Date	Patient Initials	Pre Process Volume	Pre Process Motility	Type of Process	Post Process Volume	Post Process Motility	Notes
Date	Patient Initials	Pre Process Volume	Pre Process Motility	Type of Process	Post Process Volume	Post Process Motility	Notes

Sperm Processing Cases

Date	Patient Initials	Pre Process Volume	Pre Process Motility	Type of Process	Post Process Volume	Post Process Motility	Notes

Sperm Processing Cases

Date	Patient Initials	Pre Process Volume	Pre Process Motility	Type of Process	Post Process Volume	Post Process Motility	Notes
Date	Patient Initials	Pre Process Volume	Pre Process Motility	Type of Process	Post Process Volume	Post Process Motility	

Sperm Processing Cases

Date	Patient Initials	Pre Process Volume	Pre Process Motility	Type of Process	Post Process Volume	Post Process Motility	Notes

Sperm Cryopreservation Procedure Notes

Goals and Expectations

Sperm Cryopreservation Cases

Date	Patient Initials	Pre Cryo Volume	Pre Cryo Motility	Post Cryo Motility	Post Cryo Volume	Post Cryo Motility	Post Wash Volume	Post Wash Motility	Notes

Date	Patient Initials	Pre Cryo Volume	Pre Cryo Motility	Post Cryo Motility	Post Cryo Volume	Post Cryo Motility	Post Wash Volume	Post Wash Motility	Notes

Sperm Cryopreservation Cases

Date	Patient Initials	Pre Cryo Volume	Pre Cryo Motility	Post Cryo Motility	Post Cryo Volume	Post Cryo Motility	Post Wash Volume	Post Wash Motility	Notes

| Date | Patient Initials | Pre Cryo Volume | Pre Cryo Motility | Post Cryo Motility | Post Cryo Volume | Post Cryo Motility | Post Wash Volume | Post Wash Motility | Notes |

Sperm Cryopreservation Cases

Date	Patient Initials	Pre Cryo Volume	Pre Cryo Motility	Post Cryo Motility	Post Cryo Volume	Post Cryo Motility	Post Wash Volume	Post Wash Motility	Notes

Sperm Cryopreservation Cases

Date	Patient Initials	Pre Cryo Volume	Pre Cryo Motility	Post Cryo Motility	Post Cryo Volume	Post Cryo Motility	Post Wash Volume	Post Wash Motility	Notes

Sperm Cryopreservation Cases

Date	Patient Initials	Pre Cryo Volume	Pre Cryo Motility	Post Cryo Motility	Post Cryo Volume	Post Cryo Motility	Post Wash Volume	Post Wash Motility	Notes

Sperm Cryopreservation Cases

Date	Patient Initials	Pre Cryo Volume	Pre Cryo Motility	Post Cryo Motility	Post Cryo Volume	Post Cryo Motility	Post Wash Volume	Post Wash Motility	Notes

Sperm Cryopreservation Cases

Date	Patient Initials	Pre Cryo Volume	Pre Cryo Motility	Post Cryo Motility	Post Cryo Volume	Post Cryo Motility	Post Wash Volume	Post Wash Motility	Notes
Date	Patient Initials	Pre Cryo Volume	Pre Cryo Motility	Post Cryo Motility	Post Cryo Volume	Post Cryo Motility	Post Wash Volume	Post Wash Motility	Notes

Sperm Cryopreservation Cases

Date	Patient Initials	Pre Cryo Volume	Pre Cryo Motility	Post Cryo Motility	Post Cryo Volume	Post Cryo Motility	Post Wash Volume	Post Wash Motility	Notes

Sperm Thaw Procedure Notes

Goals and Expectations

Sperm Thaw Cases

Date	Patient Initials	Time To Thaw	Post Thaw Volume	Post Thaw Motility	Post Process Volume	Post Process Motility	Notes

Sperm Thaw Cases

Date	Patient Initials	Time To Thaw	Post Thaw Volume	Post Thaw Motility	Post Process Volume	Post Process Motility	Notes

Sperm Thaw Cases

Date	Patient Initials	Time To Thaw	Post Thaw Volume	Post Thaw Motility	Post Process Volume	Post Process Motility	Notes

Sperm Thaw Cases

Date	Patient Initials	Time To Thaw	Post Thaw Volume	Post Thaw Motility	Post Process Volume	Post Process Motility	Notes

Sperm Thaw Cases

Date	Patient Initials	Time To Thaw	Post Thaw Volume	Post Thaw Motility	Post Process Volume	Post Process Motility	Notes
Date	Patient Initials	Time To Thaw	Post Thaw Volume	Post Thaw Motility	Post Process Volume	Post Process Motility	Notes

Sperm Thaw Cases

Date	Patient Initials	Time To Thaw	Post Thaw Volume	Post Thaw Motility	Post Process Volume	Post Process Motility	Notes

Sperm Thaw Cases

Date	Patient Initials	Time To Thaw	Post Thaw Volume	Post Thaw Motility	Post Process Volume	Post Process Motility	Notes

Sperm Thaw Cases

Date	Patient Initials	Time To Thaw	Post Thaw Volume	Post Thaw Motility	Post Process Volume	Post Process Motility	Notes

Insemination Procedure Notes

Goals and Expectations

Insemination Cases

Date	Patient Initials	# Of Oocytes	Sperm Quality	# Of Oocytes Inseminated	Day 1 Fertilization	Day 3 Embryo Assessment	Blastocyst Assessment	Notes

| Date | Patient Initials | # Of Oocytes | Sperm Quality | # Of Oocytes Inseminated | Day 1 Fertilization | Day 3 Embryo Assessment | Blastocyst Assessment | Notes |

Insemination Cases

Date	Patient Initials	# Of Oocytes	Sperm Quality	# Of Oocytes Inseminated	Day 1 Fertilization	Day 5 Embryo Assessment	Blastocyst Assessment	Notes
	Patient Initials	# Of Oocytes	Sperm Quality	# Of Oocytes Inseminated	Day 1 Fertilization	Day 5 Embryo Assessment	Blastocyst Assessment	Notes

Insemination Cases

Date	Patient Initials	# Of Oocytes	Sperm Quality	# Of Oocytes Inseminated	Day 1 Fertilization	Day 3 Embryo Assessment	Blastocyst Assessment	Notes

Insemination Cases

Date	Patient Initials	# Of Oocytes	Sperm Quality	# Of Oocytes Inseminated	Day 1 Fertilization	Day 5 Embryo Assessment	Blastocyst Assessment	Notes
Date	Patient Initials	# Of Oocytes	Sperm Quality	# Of Oocytes Inseminated	Day 1 Fertilization	Day 5 Embryo Assessment	Blastocyst Assessment	Notes

Insemination Cases

Date	Patient Initials	# Of Oocytes	Sperm Quality	# Of Oocytes Inseminated	Day 1 Fertilization	Day 5 Embryo Assessment	Blastocyst Assessment	Notes

Insemination Cases

Date	Patient Initials	# Of Oocytes	Sperm Quality	# Of Oocytes Inseminated	Day 1 Fertilization	Day 3 Embryo Assessment	Blastocyst Assessment	Notes

Date	Patient Initials	# Of Oocytes	Sperm Quality	# Of Oocytes Inseminated	Day 1 Fertilization	Day 3 Embryo Assessment	Blastocyst Assessment	Notes

Insemination Cases

Date	Patient Initials	# Of Oocytes	Sperm Quality	# Of Oocytes Inseminated	Day 1 Fertilization	Day 5 Embryo Assessment	Blastocyst Assessment	Notes
	Patient Initials	# Of Oocytes	Sperm Quality	# Of Oocytes Inseminated	Day 1 Fertilization	Day 5 Embryo Assessment	Blastocyst Assessment	

Insemination Cases

Date	Patient Initials	# Of Oocytes	Sperm Quality	# Of Oocytes Inseminated	Day 1 Fertilization	Day 3 Embryo Assessment	Blastocyst Assessment	Notes

ICSI Procedure Notes

Goals and Expectations

ICSI Cases

Date	Patient Initials	# Of Oocytes	Sperm Quality	# Of Oocytes ICSI'd	Day 1 Fertilization	Day 5 Embryo Assessment	Blastocyst Assessment	Notes

Date	Patient Initials	# Of Oocytes	Sperm Quality	# Of Oocytes ICSI'd	Day 1 Fertilization	Day 5 Embryo Assessment	Blastocyst Assessment	Notes

ICSI Cases

Date	Patient Initials	# Of Oocytes	Sperm Quality	# Of Oocytes ICSI'd	Day 1 Fertilization	Day 3 Embryo Assessment	Blastocyst Assessment	Notes

ICSI Cases

Date	Patient Initials	# Of Oocytes	Sperm Quality	# Of Oocytes ICSI'd	Day 1 Fertilization	Day 3 Embryo Assessment	Blastocyst Assessment	Notes

ICSI Cases

Date	Patient Initials	# Of Oocytes	Sperm Quality	# Of Oocytes ICSI'd	Day 1 Fertilization	Day 3 Embryo Assessment	Blastocyst Assessment	Notes

ICSI Cases

Date	Patient initials	# Of Oocytes	Sperm Quality	# Of Oocytes ICSI'd	Day 1 Fertilization	Day 3 Embryo Assessment	Blastocyst Assessment	Notes

ICSI Cases

Date	Patient Initials	# Of Oocytes	Sperm Quality	# Of Oocytes ICSI'd	Day 1 Fertilization	Day 3 Embryo Assessment	Blastocyst Assessment	Notes

ICSI Cases

Date	Patient Initials	# Of Oocytes	Sperm Quality	# Of Oocytes ICSI'd	Day 1 Fertilization	Day 3 Embryo Assessment	Blastocyst Assessment	Notes
Date	Patient Initials	# Of Oocytes	Sperm Quality	# Of Oocytes ICSI'd	Day 1 Fertilization	Day 3 Embryo Assessment	Blastocyst Assessment	Notes

ICSI Cases

Date	Patient Initials	# Of Oocytes	Sperm Quality	# Of Oocytes ICSI'd	Day 1 Fertilization	Day 3 Embryo Assessment	Blastocyst Assessment	Notes

Fertilization Assessment Procedure Notes

Goals and Expectations

Fertilization Assessment

Date	Patient Initials	# Of Embryos	# of 2PN's	# of 1PN's	# of 0PN's	# of 3PN's	# of DEG	Notes

Fertilization Assessment

Date	Patient Initials	# Of Embryos	# of 2PN's	# of IPN's	# of 0PN's	# of 3PN's	# of DEG	Notes
Date	Patient Initials	# Of Embryos	# of 2PN's	# of IPN's	# of 0PN's	# of 3PN's	# of DEG	Notes

Fertilization Assessment

Date	Patient Initials	# Of Embryos	# of 2PN's	# of 1PN's	# of 0PN's	# of 3PN's	# of DEG	Notes

Fertilization Assessment

Date	Patient Initials	# Of Embryos	# of 2PN's	# of 1PN's	# of 0PN's	# of 3PN's	# of DEG	Notes

Fertilization Assessment

Date	Patient Initials	# Of Embryos	# of 2PN's	# of 1PN's	# of 0PN's	# of 3PN's	# of DEG	Notes
Date	Patient Initials	# Of Embryos	# of 2PN's	# of 1PN's	# of 0PN's	# of 3PN's	# of DEG	Notes

Fertilization Assessment

Date	Patient Initials	# Of Embryos	# of 2PN's	# of 1PN's	# of 0PN's	# of 3PN's	# of DEG	Notes

Fertilization Assessment

Date	Patient Initials	# Of Embryos	# of 2PN's	# of 1PN's	# of 0PN's	# of 3PN's	# of DEG	Notes

Fertilization Assessment

Date	Patient Initials	# Of Embryos	# of 2PN's	# of 1PN's	# of 0PN's	# of 3PN's	# of DEG	Notes

Day Three Embryo Grading Procedure Notes and Assisted Hatching

Goals and Expectations

Day Three Embryo Assessment and Assisted Hatching

Date	Patient Initials	# Of Embryos	# of Cells, Cell Size, & Fragmentation Grading	AH #
Date	Patient Initials	# Of Embryos	# of Cells, Cell Size, & Fragmentation Grading	AH #

Day Three Embryo Assessment and Assisted Hatching

Date	Patient Initials	# Of Embryos	# of Cells, Cell Size, & Fragmentation Grading	AH #

Day Three Embryo Assessment and Assisted Hatching

Date	Patient Initials	# Of Embryos	# of Cells, Cell Size, & Fragmentation Grading	AH #

Day Three Embryo Assessment and Assisted Hatching

Date	Patient Initials	# Of Embryos	# of Cells, Cell Size, & Fragmentation Grading	AH #

Day Three Embryo Assessment and Assisted Hatching

Date	Patient Initials	# Of Embryos	# of Cells, Cell Size, & Fragmentation Grading	AH #

Date	Patient Initials	# Of Embryos	# of Cells, Cell Size, & Fragmentation Grading	AH #

Day Three Embryo Assessment and Assisted Hatching

Date	Patient Initials	# Of Embryos	# of Cells, Cell Size, & Fragmentation Grading	AH #
Date	Patient Initials	# Of Embryos	# of Cells, Cell Size, & Fragmentation Grading	AH #

Day Three Embryo Assessment and Assisted Hatching

Date	Patient Initials	# Of Embryos	# of Cells, Cell Size, & Fragmentation Grading	AH #

Date	Patient Initials	# Of Embryos	# of Cells, Cell Size, & Fragmentation Grading	AH #

Day Three Embryo Assessment and Assisted Hatching

Date	Patient Initials	# Of Embryos	# of Cells, Cell Size, & Fragmentation Grading	AH #

Blastocyst Embryo Grading Procedure Notes

Goals and Expectations

Blastocyst Embryo Assessment

Date	Patient Initials	# Of Embryos	Expansion Grade, Inner Cell Mass Grade, Trophectoderm Grade	Notes

Blastocyst Embryo Assessment

Date	Patient Initials	# Of Embryos	Expansion Grade, Inner Cell Mass Grade, Trophectoderm Grade	Notes

Blastocyst Embryo Assessment

Date	Patient Initials	# Of Embryos	Expansion Grade, Inner Cell Mass Grade, Trophectoderm Grade	Notes

Blastocyst Embryo Assessment

Date	Patient Initials	# Of Embryos	Expansion Grade, Inner Cell Mass Grade, Trophectoderm Grade	Notes
	Patient Initials	# Of Embryos	Expansion Grade, Inner Cell Mass Grade, Trophectoderm Grade	Notes

Blastocyst Embryo Assessment

Date	Patient Initials	# Of Embryos	Expansion Grade, Inner Cell Mass Grade, Trophectoderm Grade	Notes

Date	Patient Initials	# Of Embryos	Expansion Grade, Inner Cell Mass Grade, Trophectoderm Grade	Notes

Blastocyst Embryo Assessment

Date	Patient Initials	# Of Embryos	Expansion Grade, Inner Cell Mass Grade, Trophectoderm Grade	Notes

Blastocyst Embryo Assessment

Date	Patient Initials	# Of Embryos	Expansion Grade, Inner Cell Mass Grade, Trophectoderm Grade	Notes

Blastocyst Embryo Assessment

Date	Patient Initials	# Of Embryos	Expansion Grade, Inner Cell Mass Grade, Trophectoderm Grade	Notes
Date	Patient Initials	# Of Embryos	Expansion Grade, Inner Cell Mass Grade, Trophectoderm Grade	Notes

Embryo Transfer Procedure Notes

Goals and Expectations

Embryo Transfer

Date	Patient Initials	Fresh or Frozen	Embryo Grade	Embryo Loading	Ultrasound Transfer Status	Clear Catheter?	Pregnancy Status	Notes

Embryo Transfer

Date	Patient Initials	Fresh or Frozen	Embryo Grade	Embryo Loading	Ultrasound Transfer Status	Clear Catheter?	Pregnancy Status	Notes

Embryo Transfer

Date	Patient Initials	Fresh or Frozen	Embryo Grade	Embryo Loading	Ultrasound Transfer Status	Clear Catheter?	Pregnancy Status	Notes

Embryo Transfer

Date	Patient Initials	Fresh or Frozen	Embryo Grade	Embryo Loading	Ultrasound Transfer Status	Clear Catheter?	Pregnancy Status	Notes

Embryo Transfer

Date	Patient Initials	Fresh or Frozen	Embryo Grade	Embryo Loading	Ultrasound Transfer Status	Clear Catheter?	Pregnancy Status	Notes

Embryo Transfer

Date	Patient Initials	Fresh or Frozen	Embryo Grade	Embryo Loading	Ultrasound Transfer Status	Clear Catheter?	Pregnancy Status	Notes

Embryo Transfer

Date	Patient Initials	Fresh or Frozen	Embryo Grade	Embryo Loading	Ultrasound Transfer Status	Clear Catheter?	Pregnancy Status	Notes

Embryo Transfer

Date	Patient Initials	Fresh or Frozen	Embryo Grade	Embryo Loading	Ultrasound Transfer Status	Clear Catheter?	Pregnancy Status	Notes
Date	Patient Initials	Fresh or Frozen	Embryo Grade	Embryo Loading	Ultrasound Transfer Status	Clear Catheter?	Pregnancy Status	Notes
Date	Patient Initials	Fresh or Frozen	Embryo Grade	Embryo Loading	Ultrasound Transfer Status	Clear Catheter?	Pregnancy Status	Notes

Embryo Cryopreservation Procedure Notes

Goals and Expectations

Embryo Cryopreservation

Date	Patient Initials	Embryo Grade	Protocol	Status	Notes

Date	Patient Initials	Embryo Grade	Protocol	Status	Notes

Embryo Cryopreservation

Date	Patient Initials	Embryo Grade	Protocol	Status	Notes

Embryo Cryopreservation

Date	Patient Initials	Embryo Grade	Protocol	Status	Notes

Embryo Cryopreservation

Date	Patient Initials	Embryo Grade	Protocol	Status	Notes

Embryo Cryopreservation

Date	Patient Initials	Embryo Grade	Protocol	Status	Notes

Embryo Cryopreservation

Date	Patient Initials	Embryo Grade	Protocol	Status	Notes

Embryo Cryopreservation

Date	Patient Initials	Embryo Grade	Protocol	Status	Notes
	Patient Initials				

Embryo Cryopreservation

Date	Patient Initials	Embryo Grade	Protocol	Status	Notes
Date	Patient Initials	Embryo Grade	Protocol	Status	Notes

Embryo Thaw Procedure Notes

Goals and Expectations

Embryo Thawing

Date	Patient Initials	Embryo Grade	Protocol	Satus	Notes

Embryo Thawing

Date	Patient Initials	Embryo Grade	Protocol	Status	Notes

Embryo Thawing

Date	Patient Initials	Embryo Grade	Protocol	Status	Notes
Date	Patient Initials	Embryo Grade	Protocol	Status	Notes

Embryo Thawing

Date	Patient Initials	Embryo Grade	Protocol	Status	Notes

Embryo Thawing

Date	Patient Initials	Embryo Grade	Protocol	Status	Notes

Embryo Thawing

Date	Patient Initials	Embryo Grade	Protocol	Status	Notes

Embryo Thawing

Date	Patient Initials	Embryo Grade	Protocol	Status	Notes

Embryo Thawing

Date	Patient Initals	Embryo Grade	Protocol	Status	Notes
Date	Patient Initals	Embryo Grade	Protocol	Status	Notes

Goals and Expectations

Goals and Expectations

Goals and Expectations

Goals and Expectations

Goals and Expectations

Goals and Expectations

Goals and Expectations

Goals and Expectations

Goals and Expectations

Date	Patient Initials		Notes

Date	Patient Initials		Notes

Date	Patient initals		Notes

Date	Patient Initials		Notes
Date	Patient Initials		Notes

Date	Patient Initals		Notes

Date	Patient Initials		Notes

Date	Patient Initials		Notes
	Patient Initials		

Date	Patient Initials		Notes

Date	Patient Initials		Notes

Date	Patient Initials		Notes

WHO 5th & WHO 6th Semen Parameters

	WHO 2010	WHO 2021
Semen volume (mL)	1.5 (1.4–1.7)	1.4 (1.3–1.5)
Total sperm number (10⁶ per ejaculate)	39 (33–46)	39 (35–40)
Total motility (%)	40 (38–42)	42 (40–43)
Progressive motility (%)	32 (31–34)	30 (29–31)
Non-progressive motility (%)	1	1 (1–1)
Immotile sperm (%)	22	20 (19–20)
Vitality (%)	58 (55–63)	54 (50–56)
Normal forms (%)	4 (3–4)	4 (3.9–4)

Oocyte Maturation

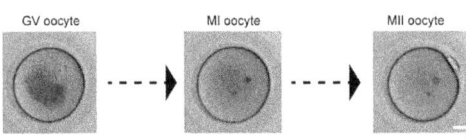

GV oocyte → MI oocyte → MII oocyte

Fertilization Assessment

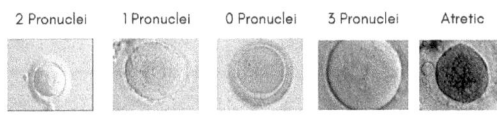

2 Pronuclei | 1 Pronuclei | 0 Pronuclei | 3 Pronuclei | Atretic

Day Three Embryo Grading

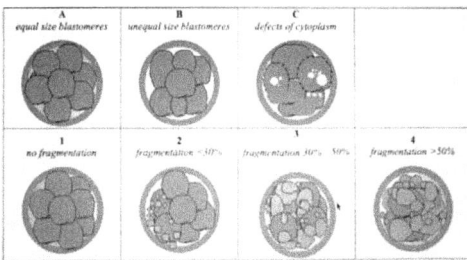

A equal size blastomeres	B unequal size blastomeres	C defects of cytoplasm	
1 no fragmentation	2 fragmentation <30%	3 fragmentation 30%–50%	4 fragmentation >50%

Gardner Embryo Grading System

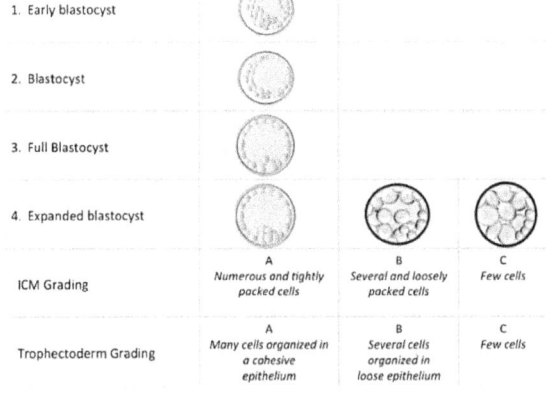

1. Early blastocyst			
2. Blastocyst			
3. Full Blastocyst			
4. Expanded blastocyst			
ICM Grading	A Numerous and tightly packed cells	B Several and loosely packed cells	C Few cells
Trophectoderm Grading	A Many cells organized in a cohesive epithelium	B Several cells organized in loose epithelium	C Few cells

www.ingramcontent.com/pod-product-compliance
Lightning Source LLC
Chambersburg PA
CBHW060846220526
45466CB00003B/1260